Project Management Practices in Business Context

The implementation of a metallurgical accounting system

Project Management Practices in Business Context

The implementation of a metallurgical accounting system

Roger Rumbu

2RA-Edition

Rumbu, Roger

Project Management Practices in Business Context - The implementation of a metallurgical accounting system / Author Rumbu, Roger

Includes bibliographical references

ISBN 978-1542971669

2017 edition

Cover: R. K. Creative Design

First edition in 2014

Project Management Practices in Business Context. ISBN: 978-1-312-12050-1

New edition by 2RA-Publishing

January 2017 in paperback.

ISBN: 978-1541066359

Cover design: R.K. Creative Design

Copyright:© Roger Rumbu, 2017

From the same author in Expertise Metallurgy series:

Extractive Metallurgy of Cobalt, 2RA-Publishing, Cape Town – South Africa, 2016.
ISBN : 978-1516843527.

Non-ferrous Extractive Metallurgy – Industrial Practices, 2RA-Publishing, Cape Town – South Africa, 2014.
ISBN : 978-1-920600-03-7.

Hydrométallurgie du cuivre - Grillage – Lixiviation – SX – Electro-extraction, 2RA-Publishing, Cape Town – South Africa, 2016.
ISBN : 978-1512138535.

Métallurgies du Zinc et des Métaux Associés, 2RA-Publishing, Cape Town – South Africa, 2016.
ISBN: 978-1516818556.

Introduction à la métallurgie extractive des terres rares, 2RA-Publishing, Cape Town – South Africa, 2012.
ISBN : 978-1-920600-28-0.

Métallurgie extractive du cobalt, 2RA-Publishing, Cape Town – South Africa, 2012.
ISBN : 978-1-920600-30-3.

Métallurgie Extractive des Non-Ferreux – Pratiques Industrielles, 3rd Edition, 2RA-Publishing, Cape Town – South Africa, 2015.
ISBN : 978-1515316299.

Métallurgie Extractive des Non-Ferreux – Pratiques Industrielles, 2nd Edition, 2RA-Publishing, Cape Town – South Africa, 2012.
ISBN : 978-1-920600-02-0.

Métallurgie Extractive des Non-Ferreux – Pratiques Industrielles, New Voices Publishing, Cape Town – South Africa, 2010.

To my lovely Mira Losa,

To our kids Andy-Grâce, Sacha-Romy, Reggie-John, Crissy-Roy and Dany-Val for their great support and patience.

To all those who contributed in one way or another to this new work completion, to you my many interested readers, in mining industry, professors, experts, scientists, engineers, my students-engineers, and technicians, passionates, friends, colleagues and my family, I will not have enough words to thank you.

Roger K.M. RUMBU

Table of Contents

Metallurgical Accounting System Implementation

Step 1: Background of the project

Ambatovy is the biggest green field nickel-cobalt metallurgical complex built by The Canadian Sherritt International Corporation on the shore of Indian Ocean in Toamasina-Madagascar (east coast) for which hot commissioning should be completed by end 2012.

Early construction work started in 2007 but final construction work and hot commissioning were completed by mid-2012.

Madagascar doesn't have neither a real mining or metallurgical culture nor an industrial one, so it was hard to find skilled local people with experience ready to operate in this up-to-date metallurgical complex.

For the start-up of a new processing plant, a metallurgical accounting system to be used as a monitoring tool should be established. The metallurgical accounting system must be an elaborated software based system.

The project we had to manage was to establish a metallurgical accounting system starting from scratch for a metallurgical plant. The main deliverable for this project was to collect all operational data to issue a Daily/Monthly/Annual reporting system for the company and Ambatovy's Top Management before operations start-up.

The metallurgical accounting system can start as an Excel based system that may be used in the future as a back-up but must move very quickly to a well-established and dedicated software based system which can operate with the company's IT infrastructure and can be operate from a cloud.

Data should be collected from metallurgical operations and all other operational activities linked to operations like power and water consumption/production, sulphuric acid and H_2S production, steam and hydrogen production and major reagents consumption like sulphur, limestone and coal.

Step 2:   Scope statement

The project we had to manage was to establish a metallurgical accounting system starting from scratch for a metallurgical plant. The main deliverable for this project was to collect all operational data to issue a Daily/Monthly/Annual reporting system for the company and Ambatovy's Top Management before operations start-up.

The metallurgical accounting system can start as an Excel based system that may be used in the future as a back-up but must move very quickly to a well-established and dedicated software based system which can operate with the company's IT infrastructure and can be operate from a cloud.

Data should be collected from metallurgical operations and all other operational activities linked to operations like power and water consumption/production, sulphuric acid and H_2S production, steam and hydrogen production and major reagents consumption like sulphur, limestone and coal.

This project duration is six months but should be completed before the end of cold commissioning of the plant just after the site had been fully energized.

The cost of this project should not exceed $500,000 and the designed metallurgical accounting system should be run by 2 to 3 persons only from Ambatovy's technical services.

The scope doesn't include future software development follow-up.

Step 3:   Life-cycle phase definition

To define the life-cycle phase, according to the size of the project, we have chosed to use the Small/Medium Project Methodology (SMPM).

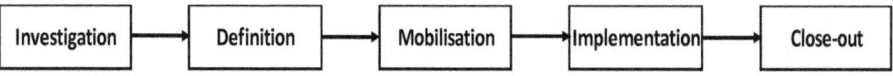

Investigation

- a) The project is to establish a metallurgical accounting system for a new process plant.
- b) The importance of this project is to be able to tune and survey operations for better efficiency.

Definition

- a) The project will be implemented from the beginning to close-out.
- b) All stakeholder are defined in step 6
- c) The project will be implemented prior operations start.
- d) The cost of the project will be of $500,000.

Mobilisation

Time for the contractor to be on site ready to accomplish the work as per the contract.

Implementation

- a) Implementation will be done by phases.
- b) In case the project is not completed by operations start-up, an interim plan of spread-sheet based metallurgical accounting system will used.

Close-out

a) Project originator and final user will review the out-come of the metallurgical accounting system.
b) All developers will sign-off their deliverables.
c) The project will be handed-over after successful cold and hot commissioning.
d) The project is the closed.

Step 4:   Work breakdown structure

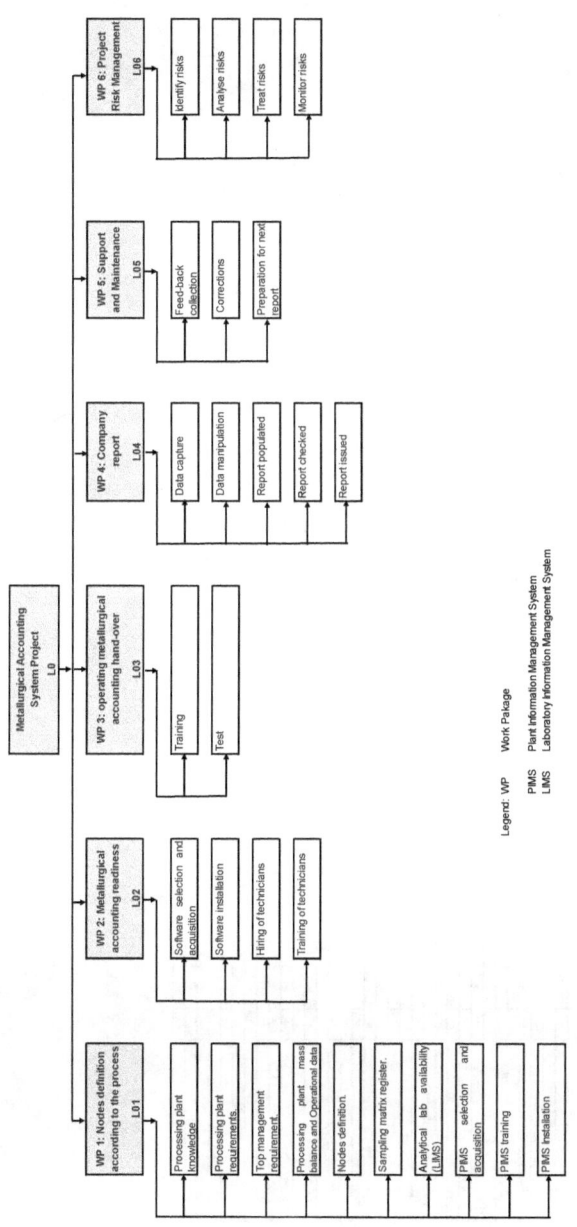

Metallurgical Accounting System Project — L0

WP 1: Nodes definition according to the process — L01
- Processing plant knowledge.
- Processing plant requirements.
- Top management requirement.
- Processing plant mass balance and Operational data
- Nodes definition.
- Sampling matrix register.
- Analytical lab availability (LIMS)
- PMS selection and acquisition
- PMS training
- PMS installation

WP 2: Metallurgical accounting readiness — L02
- Software selection and acquisition
- Software installation
- Hiring of technicians
- Training of technicians

WP 3: operating metallurgical accounting hand-over — L03
- Training
- Test

WP 4: Company report — L04
- Data capture
- Data manipulation
- Report populated
- Report checked
- Report issued

WP 5: Support and Maintenance — L05
- Feed-back collection
- Corrections
- Preparation for next report

WP 6: Project Risk Management — L06
- Identify risks
- Analyse risks
- Treat risks
- Monitor risks

Legend: WP   Work Pakage
         PMS   Plant Information Management System
         LIMS Laboratory Information Management System

7

Step 5:   Schedule

Gantt chart

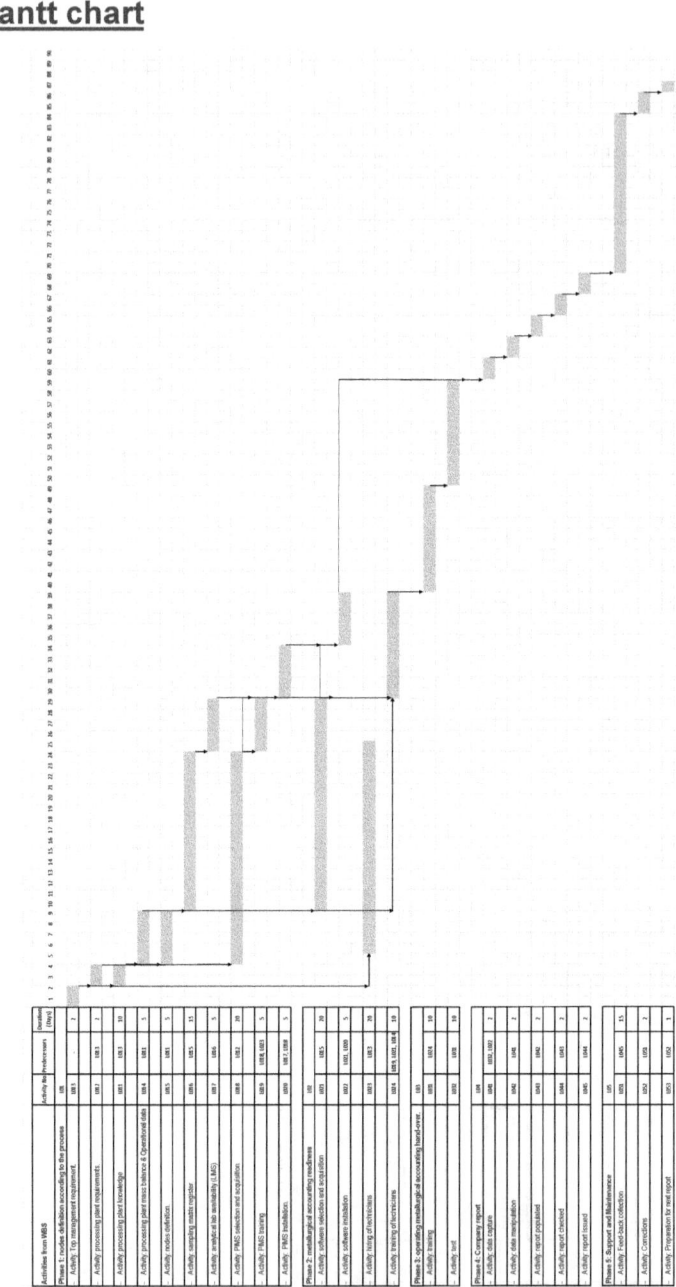

Detailed schedule

Phase 1: nodes definition according to the process

- Activity: processing plant knowledge
- Activity: processing plant requirements.
- Activity: Top management requirement.
- Activity: processing plant mass balance & Operational data.
- Activity: nodes definition.
- Activity: sampling matrix register.
- Activity: analytical lab availability (LIMS)
- Activity: PIMS selection
- Activity: PIMS training
- Activity: PIMS acquisition and installation.

Phase 2: metallurgical accounting readiness

- Activity: software selection and acquisition
- Activity: software installation
- Activity: hiring of technicians
- Activity: training of technicians

Phase 3: operating metallurgical accounting hand-over.

- Activity: training
- Activity: test

Phase 4: Company report

- Activity: data capture
- Activity: data manipulation
- Activity: report populated
- Activity: report checked
- Activity: report issued

Phase 5: Support and Maintenance

Activity: Feed-back received

- Activity: Corrections
- Activity: Preparation for next report

Step 6: Resources allocation

Stakeholders register

Category	Internal Stakehoders	Role / Needs & Expectations	Priority Influence	Interest
Owner	Ambatovy President	Real end user of the project deliverable	HP	HI
Clients	Finance Department	User of the project deliverable	LP	HI
	Operation Managers	User of the project deliverable	HP	HI
Project Sponsor	Technical Service Manager	Person that finances the project and to who the project team reports and who is organizing project working environnment	HP	HI
Functional Manager	General Manager Operations	The person who makes the project happen and receiving reports from the sponsor.	HP	HI
End users	Technical Services Engineers	The person who will operate the facility for the owner when the project is completed	LP	HI
Supporters	IT Department	The parties who provide suport, services to enable the project to be completed.	HP	LI
	HR Department	Work-force supplier and training provider	LP	LI
External Stakehoders				
Subcontractors	Algosys	Management Software Provider	LP	HI
	OSISOFT	Plant Information Management System Provider	LP	HI
	Analytical Lab Department	Supply Labs Data	LP	LI
	Support Departments	Supply Management Data	LP	LI

Algosys (Process data management company)

OSISOFT (Process data management company)

Support Departments: Supply Chain,
Maintenance and Power Plant Departments

Resources allocation

Activities from WBS	WBS No	PM Leader	Ambatovy President	Finance Department	Operation Managers	Technical Service Manager	General Manager Operations	Technical Services Engineers	IT Department	HR Department	Algosys	OSISOFT	Analytical Lab Department	Support Departments
						Internal Stakeholders						External Stakeholders		
Phase 1: nodes definition according to the process	L01													
– Activity: processing plant knowledge	L011	R			C	C	C	I		I	C	C		C
– Activity: processing plant requirements.	L012	R				C	C							
– Activity: Top management requirement.	L013		R	C	C	A	C	C			I	I		
– Activity: processing plant mass balance & Operational data	L014					A	R	C						
– Activity: nodes definition.	L015	R	I			A	C							
– Activity: sampling matrix register.	L016	R				A							C	
– Activity: analytical lab availability (LIMS)	L017					A					I	I	R	
– Activity: PIMS selection and acquisition	L018			C		R			C		C			
– Activity: PIMS training	L019					A		C	I	R		C		
– Activity: PIMS Installation.	L020					I				R			C	I
Phase 2: metallurgical accounting readiness	L02													
– Activity: software selection and acquisition	L021			C		R			C		C			
– Activity: software installation	L022					I			R		C	C		
– Activity: hiring of technicians	L023			I		C				R				
– Activity: training of technicians	L024					C				R	C			
Phase 3: operating metallurgical accounting hand-over.	L03													
– Activity: training	L031	R				C				C				
– Activity: test	L032	R	I	I	C	C	C	I	C					
Phase 4: Company report	L04													
– Activity: data capture	L041	R				C		I						
– Activity: data manipulation	L042	R				C		I						
– Activity: report populated	L043	R				C		I						
– Activity: report checked	L044	C			I	R	I	C						
– Activity: report issued	L045	C	I		I	C	I	R						
Phase 5: Support and Maintenance	L05													
– Activity: Feed-back collection	L051					R								
– Activity: Corrections	L052	C				C		R						
– Activity: Preparation for next report	L053	C				C		R						

Key:

R: Primary responsibility

A: Approve

C: Consulted

I: Informed

Step 7: Cost estimate integration to WBS

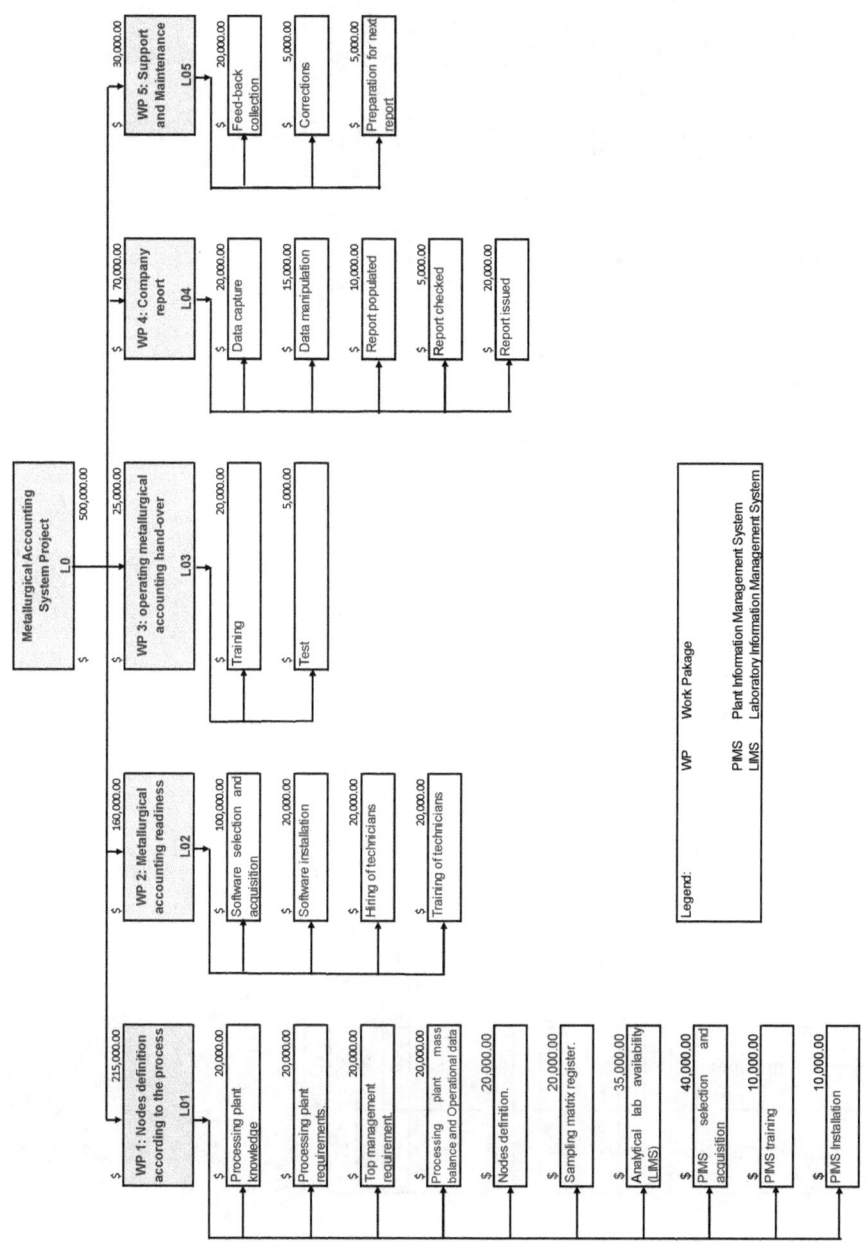

Legend:

WP	Work Pakage
PMS	Plant Information Management System
LIMS	Laboratory Information Management System

Step 8: List of key quality parameters

Quality ID	Product ID	Product	Quality Method	Producer	Reviewer(s)	Approver(s)	Target Review Date	Actual Review Date	Target Approval Date	Actual Approval Date	Result	Quality Record
1	11	Requests	Inspection	Henry	Andy	Soen	20/01	21/01	23/01	24/01	Pass	QR 11
2	21	Flow-sheet	Inspection	Henry	Andy	Soen	15/01	15/01	23/01	24/01	Pass	QR 21
3	31	Sample Register	Inspection	Henry	Andy	Soen	10/01	10/01	23/01	24/01	Pass	QR 31
4	41	Mass Balance	Inspection	Henry	Andy	Soen	10/01	10/01	23/01	24/01	Pass	QR 41
5	51	LIMS	Data validation	Henry	Andy	Soen	10/01	10/01	23/01	24/01	Pass	QR 51
6	61	PIMS	Data validation	Henry	Andy	Soen	10/01	10/01	23/01	24/01	Pass	QR 61
7	72	Main Software	Inspection	David	Andy	Soen	10/01	10/01	23/01	24/01	Pass	QR 72
8	82	Historian	Data validation	David	Andy	Soen	10/01	10/01	23/01	24/01	Pass	QR 82
9	92	Metallurgical Accounting Software	Inspection	David	Andy	Soen	10/01	10/01	23/01	24/01	Pass	QR 92
10	103	Template	Inspection	Bill	Andy	Soen	15/01	15/01	23/01	24/01	Pass	QR 103
11	113	Records	Data validation	Bill	Andy	Soen	20/01	21/01	23/01	24/01	Pass	QR 113
12	123	Data Control	Data validation	Bill	Andy	Soen	22/01	22/01	23/01	24/01	Pass	QR 123
13	133	Company Report	Inspection	Bill	Andy	Soen	23/01	24/01	23/01	24/01	Pass	QR 133

Step 9: Risk matrix and plan development

1. Risks identification

Risk No.	Risk Event / Condition Description
1	Poor knowledge of the process and the plant
2	Database infection by viruses
3	Improper data capture
4	LIMS software compatibility
5	PIMS software compatibility
6	Data recovery failure from historian
7	Metallurgical Accounting Systeme Software availability
8	Main software availability
9	Input/Outup equipment availability (Hardware)
10	Met Accounting Technicians availability
11	Improper management requets
12	Improper measurement tools
13	Poor data manipulation knowledge
14	Volontary insane data manipulation by operators or third party
15	Delay in repport delivery
16	Low report consistency
17	Higher priority projects
18	Systems/infrastructure upgrades
19	Scope change
20	Unexpected coding

2. <u>Risks analysis</u>

Risk No.	Risk Event / Condition Description	P	Total Loss (R)	Expected Loss (R)
1	Poor knowledge of the process and the plant	0.10	1,000,000	100,000
2	Database infection by viruses	0.30	2,000,000	600,000
3	Improper data capture	0.20	500,000	100,000
4	LIMS software compatibility	0.10	500,000	50,000
5	PIMS software compatibility	0.10	500,000	50,000
6	Data recovery failure from historian	0.10	2,000,000	200,000
7	Metallurgical Accounting Systeme Software availability	0.20	100,000	20,000
8	Main software availability	0.05	2,000,000	100,000
9	Input/Outup equipment availability (Hardware)	0.20	2,000,000	400,000
10	Met Accounting Technicians availability	0.40	2,000,000	800,000
11	Improper management requets	0.30	100,000	30,000
12	Improper measurement tools	0.10	2,000,000	200,000
13	Poor data manipulation knowledge	0.30	1,000,000	300,000
14	Volontary insane data manipulation by operators or third party	0.10	2,000,000	200,000
15	Delay in repport delivery	0.30	2,000,000	600,000
16	Low report consistency	0.30	1,000,000	300,000
17	Higher priority projects	0.30	450,000	135,000
18	Systems/infrastructure upgrades	0.60	160,000	96,000
19	Scope change	0.30	200,000	60,000
20	Unexpected coding	0.10	150,000	15,000

Risk No.	Risk Event / Condition Description	P	Total Loss (R)	Expected Loss (R)	Cumulative
10	Met Accounting Technicians availability	0.40	2,000,000	800,000	800,000
2	Database infection by viruses	0.30	2,000,000	600,000	1,400,000
15	Delay in repport delivery	0.30	2,000,000	600,000	2,000,000
9	Input/Outup equipment availability (Hardware)	0.20	2,000,000	400,000	2,400,000
13	Poor data manipulation knowledge	0.30	1,000,000	300,000	2,700,000
16	Low report consistency	0.30	1,000,000	300,000	3,000,000
6	Data recovery failure from historian	0.10	2,000,000	200,000	3,200,000
12	Improper measurement tools	0.10	2,000,000	200,000	3,400,000
14	Volontary insane data manipulation by operators or third party	0.10	2,000,000	200,000	3,600,000
17	Higher priority projects	0.30	450,000	135,000	3,735,000
1	Poor knowledge of the process and the plant	0.10	1,000,000	100,000	3,835,000
3	Improper data capture	0.20	500,000	100,000	3,935,000
8	Main software availability	0.05	2,000,000	100,000	4,035,000
18	Systems/infrastructure upgrades	0.60	160,000	96,000	4,131,000
19	Scope change	0.30	200,000	60,000	4,191,000
4	LIMS software compatibility	0.10	500,000	50,000	4,241,000
5	PIMS software compatibility	0.10	500,000	50,000	4,291,000
11	Improper management requets	0.30	100,000	30,000	4,321,000
7	Metallurgical Accounting Systeme Software availability	0.20	100,000	20,000	4,341,000
20	Unexpected coding	0.10	150,000	15,000	4,356,000

3. Risks Map and Threshold Line

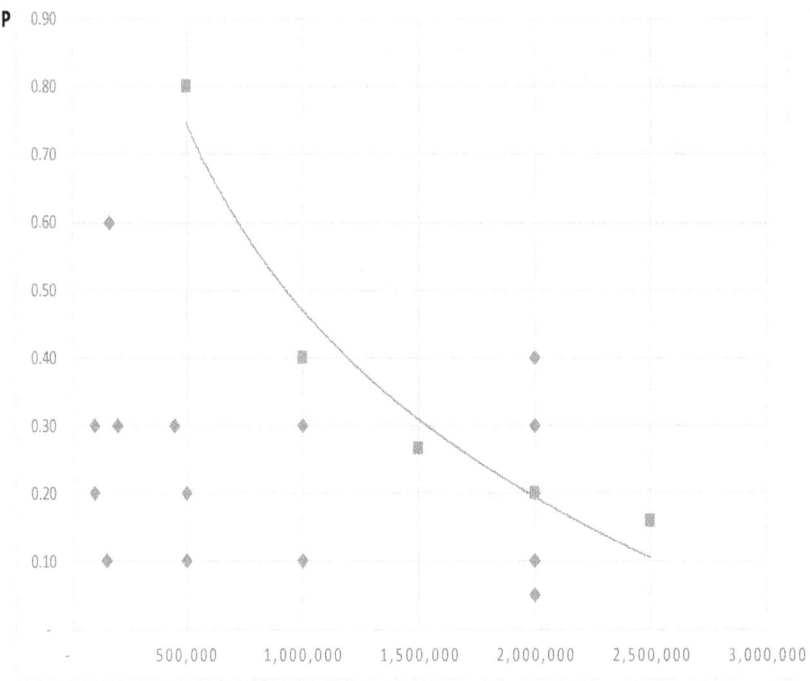

4. Critical Risks

Risk No.	Risk Event / Condition Description	P	Total Loss (R)	Expected Loss (R)
10	Met Accounting Technicians availability	0.40	2,000,000	800,000
2	Database infection by viruses	0.30	2,000,000	600,000
15	Delay in repport delivery	0.30	2,000,000	600,000
9	Input/Outup equipment availability (Hardware)	0.20	2,000,000	400,000

5. <u>Risk Response Plan for Metallurgical Accounting Project</u>

Risk No.	Risk Event / Condition Description	Risk Response
10	Met Accounting Technicians availability	Reduce the risk by start hiring processus locally as soon the project is signed. Reduce the risk by hiring temporarely expatriates met accounting technicians as contengency Reduce the risk by transfering Met accounting management to a subcontrator
2	Database infection by viruses	Reduce the risk by using redundant or splited databases Reduce the risk by using up-to-date and effective security software and procedures
15	Delay in repport delivery	Reduce the risk by choosing a suitable closure period to permit to issue the report in time Reduce the risk by twickling all reasons leading to reporting delay.
9	Input/Outup equipment availability (Hardware)	Reduce the risk by allocating enough time and resources to supply chain team and technicians to look and test apprpriate material.

6. Risk Management Monitoring

Very often, companies doesn't care about risks and find themselves in trouble when something goes wrong during the project.

According to Project Risk Management, the risk is defined as the systematic processes of identifying, analysing and responding to project risk.

Managing project risks is the ultimate in proactive management. The goal of managing project risks is to identify and prepare for any potential threat to the project's critical success factors before it actually occurs.

Care must be taken at the existence of a Project Risk Management team before project's start-up since early installation, construction, POVT (Pre Operation – Verification and Testing), cold and hot commissioning.

This team will exist only for the project and will be made of experienced people with PRM training and experience.

A subcontractor dedicated only to PRM can be hired if nobody is skilled enough for this important task.

The advantage of taking care of PRM will prevent the project about unwanted drawbacks by eliminating, avoiding or reducing negative situations leading to projects extra costs, delays or even cancellation.

Step 10: Communication

1. Meetings

Meetings	Periodicity	Objects	Attendees
Daily meeting	Daily	To give the daily advancement of the project.	PM Leader Technical Services Engineers
Weekly / Progress Meeting	Weekly	To present the plan for the following week and to talk about main issues from the last week.	PM Leader Technical Services Engineers Technical services Manager
Monthly Meeting	Monthly	To give the monthly outcome of the project.	PM Leader Technical services Manager Operation Managers Support Departments
Stakeholders Meeting	On request	To transmit main informations about the project.	Invited stakehoders

2. <u>Reportings</u>

Reports	Periodicity	Objects	Issued by	When	Destination
Daily Report	Daily	To give the daily advancement of the project.	PM Leader	The same day	Technical Services Engineers Technical services Manager
Weekly / Progress Meeting	Weekly Meeting	To present the plan for the following week and to talk about main issues from the last week.	PM Leader	Within 48 hours	Technical Services Engineers Technical services Manager
Monthly Meeting	Monthly	To give the monthly outcome of the project.	Technical services Manager	Within 48 hours	PM Leader Technical Services Engineers Operation Managers Support Departments Relevent stakeholders
Budget variance report	Monthly	Reporting of differences between planned and actual budget	PM Leader	End of Month	Technical services Manager
Stakeholders Meeting	After meeting	To transmit main informations about the project.	Technical services Manager	Within 48 hours	Invited stakehoders PM Leader

Step 11: Commercial & Procurement

1. Contracting strategies

The contracting strategy to be chosen will be The Cost Reimbursement Contract Procurement Strategy.

- Under this contractual arrangement, the client undertakes to pay the contractor, here the software supplier, the actual cost of labour, plant and material used in the execution of the work and agrees on a charge to cover the contractor's overheads and profit. This charge is normally expressed as a percentage of the actual cost.

This strategy has been chosen because:

- The client wishes to have the flexibility to influence the execution of the works and assume the entire risk.
- The extent of the work cannot be accurately predicted but early start is required.
- High quality standards apply.

2. Contracting formats

The contracting format to apply with a chosen software supplier after tender have been done is as follow.

2.1 The essentialia for the specific contract.

About the work to be done
- The client wants the contractor to provide and install a Metallurgical Accounting software.
- The software should be compatible with the client main software and database.
- The construction style will be New Renaissance style.

23

- The software should not use excessive company's bandwidth.
- The software should be provide with an efficient data recovery plan.

About the payment

The pricing to be applied is linked to the number of nodes to be surveyed. The cost per node will be vat inclusive at the rate calculate on the 1st January 2014.

The payment will be done per phase, each operating plant section constitute one phase.

The bill of material must be completed in Republic of South Africa or in Western Europe and absolutely not in Russia, China or in India.

Working hours

The weekly working hours should be 45 hours as per the law.

All work should be performed during working hours.

About the end date

After contract acceptance, the software installation activity will have a duration day of 150 calendar days excluding week-end and official public holidays.

Independence of the contractor

The contractor is independent and free of contract with the one requesting the work to be done to avoid to deal with an employee during the contract.

2.2   Any suspensive condition.

- The constructor must hold all Consent to Commence Construction document:
    - IT Safety mandatory document;

- Valid letter of good standing;
- Proof of registration issued;
- Subcontractor agreement signed.
- A client IT security manager must be part of the contractor's team.
- All workers used for the work must hold a valid work permit with the constructor.
- In case of positive malperformance: If the construction is significantly deviating from plan, the contractor will be charged.

2.3 Any resolutive time period.

- In the case of something happened to the client and if no succession plan is not presented to the contractor, the work will be stopped.
- The installation guaranty exists up to 36 (thirty six) months after work completion.

2.4 A cancellation clause.

- If the contractor loses his registration.
- If the contractor uses a fake registration.
- If the client is not able to pay the contractor the money due after 50% completion.
- If there is a default by debtor on origin of software.
- In case of positive malperformance: the construction is significantly not done according to the plan.
- In case of repudiation: when the contractor announce without explanation that he won't be able to deliver his work in time.
- Upgrading activities done without the permanent presence of the client's IT security manager.
- Lack of deviation reporting.
- If the rand value to the dollar exceed R15 (force majeure).

2.5   A penalty clause in favour of the client

- In case the work is not completed and in absence of unforeseen circumstance or force majeure after the due completion date, the contractor will be charged 1% of the value of the contract per day of delay until construction completion.
- If another contractor is chosen to complete the work after failure of the initial contractor to complete its part of the contract, they will be charged for the work completed by the new contractor.

2.6   A clause that deals with force majeure on unforeseen circumstances event.

- In case power availability is detrimental to upgrade work, the work will be stopped and will resume only on positive circumstances.

2.7   A clause on an effective way to formally amend the contract document to reflect the changes agreed to by the parties.

- When a change is required, a meeting must be requested immediately by the one requiring the change and held and the amendment will be added to the contract.

Reference

1. Dr. MC Bekker, Project Management in Organisationel Context – Study Guide, Issue 12, Jan. 2014.
2. R. Burke, Project Management Techniques, College Edition, Project Management Series, 2007, pp. 379.
3. G.M. Horine, Project Management – Absolute Beginner's Guide, Third Edition, Que Publishing, 2013, pp. 429.
4. J.J.A. Jansen, Project Procuremment Management – Study Notes, 2014.
5. B. Kuschke, General Law of Contract – Critical Issues and Risks, Legal Aspect of Project Management – Programme in Project Management, CE-University of Pretoria, 2014.
6. J.R. Meredith et al., Project Management – A Managerial Approach, 8th Edition, Wiley Ed., 2012, pp. 587.
7. R. Rumbu, Project Management Practices in Business Context, 2RA-Publishing, 2014.
8. H. Steyn et al., Project Management – A Multi-Disciplinary Approach, Third Revised Edition, FPM Publishing, 203, pp. 478.
9. K. Visser, Study guide for Project Risk Management – Programme in Project Management, CE-University of Pretoria, 2014.

Published by 2RA – PUBLISHING

January 2017

Sandton, R.S.A.

Email: edition@2ra-company.com

ISBN: **978-1542971669**

www.ingramcontent.com/pod-product-compliance
Lightning Source LLC
Chambersburg PA
CBHW070727180526
45167CB00004B/1644